THE POCKET
CLOUD BOOK

THE POCKET CLOUD BOOK

HOW TO UNDERSTAND THE SKIES

RICHARD HAMBLYN

In association with the Met Office

DAVID & CHARLES

www.davidandcharles.com

Contents

Introduction

One of my favourite *Peanuts* cartoons features Linus and Charlie Brown, lying on their backs, gazing up at the passing clouds. When Charlie asks Linus if he can see any shapes in them, Linus replies that he's just spotted the outline of British Honduras, the profile of the painter Thomas Eakins, and a remarkably detailed tableau of the stoning of St Stephen: 'there's the Apostle Paul standing to one side. What about you, Charlie Brown?': 'I was just going to say I saw a ducky and a horsey, but I just changed my mind'.

Clouds have been objects of delight and fascination throughout history, their fleeting magnificence and endless variability providing food for thought for scientists and daydreamers alike.

'For if there is nothing new on Earth, still there is something new in the heavens. We have always a resource in the skies … and the inquiring mind may always read a new truth.'

(*The Journal of Henry D. Thoreau*, ed. B. Torrey & F. H. Allen, 14 vols (New York, 1962): I, p.11)

Clouds and their classification

In contrast to all other earthly phenomena, clouds (at least in Western culture) remained uncatalogued and unnamed until the early nineteenth century, when the Latin terms that are now in international use – 'cirrus', 'stratus', 'cumulus' and their compounds – were bestowed on them by Luke Howard (1772–1864), a 30-year-old pharmacist and meteorologist from East London.

Howard devised a simple classificatory system which overcame the challenge of the clouds' continual merging and demerging, as they rise, fall and spread through the atmosphere, rarely maintaining the same shape for more than a few minutes at a time. Howard's classification was soon taken up across the scientific world, and by the early 1820s the landscape painter John Constable was using his own, heavily annotated copy of Howard's classification in connection with the sequence of more than 100 cloud studies that he painted in the open air at Hampstead Heath.

As the nineteenth century wore on, Howard's original classification was gradually refined and enlarged in accordance with new insights and observations into the behaviour of clouds and weather. During the twentieth century it developed into the current international system of cloud classification based on ten cloud genera, or types, classified by height (low, medium and high) and within three major groups (cumulus, stratus and cirrus).

Part of the ongoing success of Howard's classificatory system is that it has proved compatible with the international system of binomial nomenclature introduced by Linnaeus in the early eighteenth century, based on the organising concept of a genus subdivided into two or more species. Using the Linnean principle, most clouds can be identified as belonging to a particular species as well as to one of the ten main genera, with their specific name alluding to a characteristic shape or structure. A cirrus uncinus,

John Constable, 'Study of Cirrus Clouds', oil on paper, c.1822. Constable owned a copy of Luke Howard's cloud classification, and on the reverse of this oil sketch, he wrote the word 'cirrus'.

for example, is an easily recognised hook-shaped species of cirrus cloud, *uncinus* deriving from the Latin word for 'hooked'.

A further subdivision relates to cloud varieties, used to identify certain additional characteristics, such as a cloud's transparency, or a particular arrangement of its elements. Altocumulus stratiformis duplicatus, for example, refers to a stratiform species of altocumulus that occurs in multiple layers with *duplicatus*, the Latin word for 'doubled', referring to an additional characteristic.

The chart on pages 10–11 shows how clouds are organised according to height, the genera of clouds that occur within each altitude band and the species found within each of the genera. After each species, in brackets, is the international code for that cloud. The code for each species is made up of the letter 'C' for cloud, 'L' (low), 'M' (medium) or 'H' (high) denoting the altitude band for that cloud and, finally, the number assigned to the cloud on a scale of 1–9. Also listed are the varieties that occur within each of the genera.

Cloud classification chart

LOW CLOUDS

Cumulus (Cu)

humilis
fractus
congestus
mediocris
radiatus
flammagenitus
homogenitus

Cumulonimbus (Cb)

calvus
capillatus

Stratocumulus (Sc)

stratiformis
lenticularis
castellanus
cumulogenitus
floccus
volutus
translucidus
perlucidus
opacus
duplicatus
undulatus
radiatus
lacunosus

Stratus (St)

nebulosus
fractus
opacus
translucidus
undulatus
cataractagenitus

MEDIUM CLOUDS

Altostratus (As)

translucidus
opacus
duplicatus
undulatus
radiatus

Nimbostratus (Ns)

Altocumulus (Ac)

stratiformis
lenticularis
castellanus
floccus
volutus
translucidus
perlucidus
opacus
duplicatus
undulatus
radiatus
lacunosus

HIGH CLOUDS

Cirrus (Ci)

fibratus
uncinus
spissatus
castellanus
floccus
intortus
radiatus
vertebratus
duplicatus
homogenitus

Cirrostratus (Cs)

fibratus
nebulosus
duplicatus
undulatus

Cirrocumulus (Cc)

stratiformis
lenticularis
castellanus
floccus
undulatus
lacunosus

CLOUD VARIETIES

duplicatus	more than one layer at different levels
intortus	irregular or tangled
lacunosus	thin cloud with regularly spaced holes, net-like
opacus	completely masks sun or moon
perlucidus	broad patches with some (small) gaps allowing blue sky to be seen
radiatus	broad parallel bands converging owing to perspective
translucidus	translucent enough to permit the sun or moon to be seen
undulatus	sheets with parallel undulations
vertebratus	resembling ribs or bones

How to use this book

This book is intended to enable you not only to identify individual clouds and skies as they might appear at any given moment, but also to track their likely changes over time. Cloud behaviour – their capacity to transfer allegiance from one modification to another within the space of a few minutes or hours – is an integral part of cloud classification. The thumbnail images on the following pages provide a quick reference point for the initial identification of a cloud or type of sky.

The clouds are listed in the order of the unique code identification system from the *International Cloud Atlas*. The principal advantage of employing this classificatory scheme is that it places a much greater emphasis on the processes of cloud growth and decay that determine the ever-changing appearance of our skies. Many of these individual classifications describe changing states of the sky as a whole, often tracing the eventful careers of certain cloud species over fairly lengthy periods of time. Also given are the unique symbol for each species and its meteorological specification.

In addition to the ten principal clouds there are four accessory clouds: pileus ('cap cloud'), pannus ('scud'), velum ('veil cloud'), and flumen ('flow cloud'), which occur only in conjunction with the principal types, as well as eleven supplementary features: arcus ('arch'), asperitas ('roughened cloud'), cauda ('tail'), cavum ('fallstreak hole'), fluctus ('wave cloud'), incus ('anvil'), mamma ('udders'), murus ('wall cloud'), praecipitatio ('rain'), tuba ('funnel') and virga ('fallstreaks'), most of which are occasional players in the riotous displays put on by large cumulonimbus clouds.

Cloud Classification Symbols

Each of the 27 cloud states has been assigned a unique international symbol, which is used by meteorologists and aviators as a form of rapid visual shorthand when reporting the state of the sky.

LOW	MEDIUM	HIGH
C_L1	C_M1	C_H1
C_L2	C_M2	C_H2
C_L3	C_M3	C_H3
C_L4	C_M4	C_H4
C_L5	C_M5	C_H5
C_L6	C_M6	C_H6
C_L7	C_M7	C_H7
C_L8	C_M8	C_H8
C_L9	C_M9	C_H9

Quick reference

C_L1 (page 16)
C_L2 (page 20)
C_L2 (page 21)
C_L3 (page 27)
C_L4 (page 29)
C_L5 (page 32)
C_L5 (page 33)
C_L5 (page 33)
C_L6 (page 38)
C_L6 (page 41)
C_L7 (page 42)
C_L8 (page 44)
C_L9 (page 46)
C_M1 (page 50)
C_M2 (page 52)
C_M3 (page 54)
C_M3 (page 55)
C_M4 (page 56)

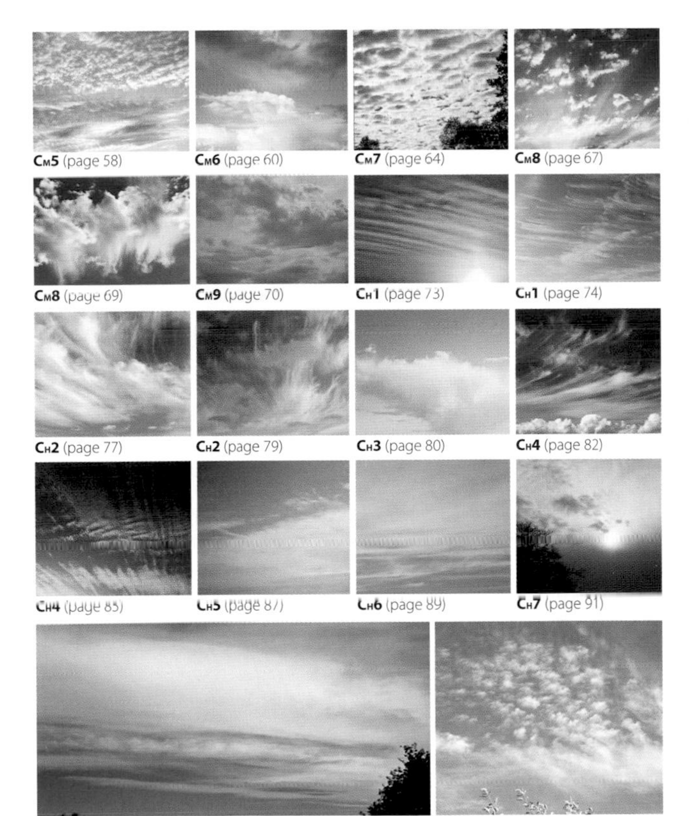

C_M5 (page 58)

C_M6 (page 60)

C_M7 (page 64)

C_M8 (page 67)

C_M8 (page 69)

C_M9 (page 70)

C_H1 (page 73)

C_H1 (page 74)

C_H2 (page 77)

C_H2 (page 79)

C_H3 (page 80)

C_H4 (page 82)

C_H4 (page 85)

C_H5 (page 87)

C_H6 (page 89)

C_H7 (page 91)

C_H8 (page 92)

C_H9 (page 96)

$C_L 1$

SPECIFICATION: **Cumulus clouds with little vertical extent and seemingly flattened (humilis), or ragged cumulus other than that of bad weather (fractus).**

Symbol = ◠

CUMULUS FRACTUS & CUMULUS HUMILIS

Appearance and nature

These convective clouds are formed above thermals (columns of ascending air) that rise in plumes from the sun-warmed ground, with the smaller cumulus fractus clouds sometimes seen emerging from fragments of haze on warm, calm summer mornings. Seeding themselves on condensation nuclei – microscopic grains of dust, smoke, pollen or sea-salt that are naturally present in the air – the rising pockets of water vapour cool and condense into droplets, which then begin to coalesce, growing upwards and outwards into a puffy white cloud.

Cumulus fractus (Cu fra) clouds at an early stage of their formation, rising on thermals from the sun-warmed ground.

A quiet summer sky filled with cumulus humilis (Cu hum) over New South Wales, Australia.

Weather implications

Over land, on clear mornings, cloudlets of cumulus may form as the sun rapidly heats the ground, or may result from the transformation of patches of foggy stratus nebulosus (see C_L6). If these smaller cumulus formations begin to show moderate vertical development, especially on warmer afternoons after rain, when the atmosphere may be growing a little unstable, they begin to be classified as cumulus mediocris or cumulus congestus (see C_L2), the main difference between this trio of species being that cumulus humilis clouds are wider than they are tall, while cumulus mediocris clouds are as tall as they are wide, and cumulus congestus clouds are even taller. But not all C_L1 clouds are destined to grow into C_L2 species, and at the end of calm summer days when the warm air begins to cool at sunset, and the thermals cease their rising, both these cumulus species will begin to sink and dissipate, breaking down into ever smaller fragments. Since neither of these smaller cumulus species are rain-bearing clouds, their evening disappearing act is due to changes in air temperature, not to any shrinkage through precipitation.

Modifications and transitions

When completely formed, these dense, white, detached clouds, with wide areas of blue sky between them, have clear-cut horizontal bases and rounded tops, and are often referred to as 'fair-weather' cumulus, particularly if they rise and consolidate to become cumulus humilis clouds, as in the photograph on page 17.

Cumulus fractus clouds beginning to dissipate as evening falls over San Francisco Bay.

C_L2

CUMULUS CONGESTUS & CUMULUS MEDIOCRIS

SPECIFICATION: Cumulus clouds of moderate or strong vertical extent, generally with protuberances in the form of domes or towers, sometimes accompanied by other cumuliform clouds, all with their bases at the same level.

Symbol = 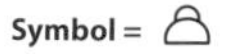

Appearance and nature

These larger cumulus clouds are developments of C_L1 clouds, formed by the upward convection of columns of warm, moist air on sunlit days. As these thermals rise,

Puffy white cumulus congestus (Cu con) clouds grow rapidly through upward convection on sunlit days.

they expand and cool until they reach the dew point (the temperature at which water vapour condenses into droplets), at which point their payload of moisture condenses and coalesces into clouds. The condensation process releases a great deal of latent heat (thermal energy locked up in water vapour), which serves to warm the air inside the growing cloud, leading to further buoyant convection and thus to further build-up of the cloud. The more this cycle continues, the taller the cloud will grow, especially if the process begins early in the morning, with a full day of sunshine ahead of it.

Cumulus mediocris (Cu med) clouds, arranged into parallel lines known as 'cloud streets'.

Modifications and transitions

The cumuliform cycle depends to a great extent on the stability and temperature of the surrounding air: if the rising moisture matches the temperature of warm, stable surrounding air, it will tend to spread out, into stratiform clouds, rather than grow upwards; but if the rising moisture remains surrounded by cooler air, it will carry on rising to form cumuliform clouds, a sure sign of atmospheric instability.

There are plenty of external limits to cumulus cloud growth, however, such as wind shear from the side, evaporation from above, and uneven convection from below (varied surfaces on the ground reflect varying amounts of heat radiation), so that some cumulus clouds will grow only to the size seen in the photographs on p. 20–21.

Weather implications

With a day's worth of uplift under their belts, however, energetic cumulus clouds can grow extremely large, as can be seen in the photograph opposite, their great white convective turrets rearing a mile or more into the sky. Over land, such clouds may well have begun to disperse by the early evening, as thermal convection currents rapidly diminish, but over the oceans, as in this case, cumulus growth often carries on late into the night, as the sea gives up its absorbed heat radiation over much longer periods. If formed at sea, cumulus clouds of this size (more than 2km/6,500ft) high,

A vigorous cumulus congestus formation.

can go on to produce light showers, and sometimes even heavy rainfall in the tropics, with larger cumulus congestus often graduating either upwards into cumulonimbus calvus clouds (see following entry, C_L3), or outwards, due to horizontal spreading at a temperature inversion, to form thick layers of stratocumulus cumulogenitus (see C_L4).

C_L3

SPECIFICATION: **Cumulonimbus cloud, its summit lacking sharp outlines, being neither clearly fibrous, nor in the shape of an anvil.**

Symbol =

CUMULONIMBUS CALVUS

Appearance and nature

Vast, roiling structures whose summits approach the upper limits of the troposphere, cumulonimbus calvus clouds represent the next stage of development of large cumulus congestus clouds. As the congestus clouds continue to build, their tops begin to lose the clearly defined 'cauliflower' appearance, becoming smoother and more consolidated, which, for cloud identification purposes, is the moment when a cumulus congestus (C_L2) classification gives way to cumulus calvus (C_L3), calvus deriving from the Latin word for 'bald'. This moment of transition from a C_L2 to a C_L3 is

captured in this photograph of an ergetic young cumulonimbus calvus. It is clearly still growing fast, as indicated by the vigorous

A cumulonimbus calvus cloud (Cb cal) in a later stage of development, its summit swirling with vigorous convective energy.

protuberances, and the blue sky behind that shows that there is still plenty of solar power at its disposal.

From a distance it is easy enough to distinguish a cumulonimbus calvus from its stormier relation, the cumulonimbus capillatus, although it is another matter if one happens to be immediately beneath it.

Modifications and transitions

These clouds can grow to considerable heights, as can be seen in the photograph on the previous page, of a large cumulonimbus calvus, the summit of which still appears to be swirling with convective energy, although it is also beginning to spread outwards, suggesting that it is in yet another transitional phase, and that this cloud is rapidly on the way to developing into a full-blown stormcloud, complete with an icy storm-brewing thunderhead that will form from the freezing of the summit into clearly visible wind-blown striations (see cumulonimbus capillatus, C_L9).

Weather implications

Cb cal clouds can produce heavy rain and squalls, but they rarely produce lightning or hail, which tend to be the unique province of Cb cap clouds.

Both species of cumulonimbus cloud tend to be given a wide berth by aircraft, since strong currents within them create powerful turbulence, while their high water content can also result in thick layers of ice forming on the cold metal.

The moment of transition from a cumulus congestus to a cumulonimbus calvus (Cb cal) is captured here.

C_L4

Stratocumulus cumulogenitus in a late afternoon sky.

SPECIFICATION: **Stratocumulus formed from the spreading out of cumulus clouds, the remains of which may also be apparent in the sky.**

Symbol = ⌓

STRATOCUMULUS CUMULOGENITUS

Appearance and nature

Stratocumulus cumulogenitus (Sc cugen) is formed when rising cumulus congestus clouds begin to spread out horizontally, as can be seen in the image opposite of a late afternoon sky. A prime example of a cloud in transition, as it changes from one genera to another, this species of stratocumulus occurs when the upper parts of rising cumulus congestus clouds (see C_L2) encounter what is known as a temperature inversion – a layer of air in which the normal rate of atmospheric cooling has either slowed down or gone into reverse, due to the presence of warm air currents moving at higher altitudes. On the whole, rising thermals are not strong enough to punch their way through a temperature inversion, although sometimes some

isolated patches of cumulus growth do resume above the flat-topped layer of stratocumulus (see also the sky state described in C_L8). Upon meeting this barrier to upward convection (air is a very poor conductor of heat), the rising cloud begins to spread horizontally instead, creating a characteristic tapered appearance. After a while, the separate bases of the cumulus clouds link up, creating an extensive cloud structure that can cover a large area of sky.

Modifications and transitions

As the warmth of the sun decreases at evening, cumulus clouds often flatten into patches of stratocumulus, a process that can be seen in this sequence of four images taken over a 20-minute period. Patches of cirrus and cirrostratus clouds also appear in these photographs.

Weather implications

These clouds spread out at the base of a temperature inversion; should they thicken and spread to cover a large area of sky, they could produce a small amount of precipitation, although this would tend to be short-lived drizzle, since this low cloud variety is formed by an impediment to the kind of upward convection from which heavier storm-clouds develop.

3

4

C_L5 STRATOCUMULUS STRATIFORMIS, STRATOCUMULUS CASTELLANUS, STRATOCUMULUS LENTICULARIS, STRATOCUMULUS VOLUTUS

SPECIFICATION: Stratocumulus not resulting from the spreading out of cumulus clouds.

Symbol = ⁓

Appearance and nature

Stratocumulus clouds are the most common clouds on Earth, being routinely visible over vast tracts of land and sea. Stratocumulus stratiformis tend to form from the lifting or breaking up of low sheets of stratus (C_L6) by upwards convection, which thicken into dark rolls, their edges merging to form an apparently continuous layer of cloud; this can be seen in the photograph below.

Stratocumulus stratiformis (Sc str) form when layers of low-lying stratus rise and thicken into dark, rolling layers of cloud.

Turrets of stratocumulus castellanus (Sc cas) rise from a horizontal base, sometimes growing upwards to develop into cumulus congestus clouds.

Stratocumulus lenticularis (Sc len) clouds form in layers of air rising over hills, as in this example of a cloud-filled afternoon sky above the German town of Limburg.

Stratocumulus castellanus features convective cumuliform turrets rising from a horizontal base, as can be seen in the photograph on page 33. If convection continues, these turrets may grow upwards over the course of a day, developing into cumulus congestus (C_L2), or even cumulonimbus clouds, and from there, perhaps, into stratocumulus cumulogenitus, coming full circle from Sc to Cu and back to Sc.

Stratocumulus lenticularis is one of the less common species in the Sc genus. It forms in moist air that rises gently over hills, creating long, smooth, lenticular shapes, as can be seen in the accompanying photograph. This kind of wave

Stratocumulus volutus ('roll cloud') off the coast of Punta del Este, Uruguay.

cloud looks very different from altocumulus lenticularis (see C_M4), being more undulating, and extended in length, and thus not so UFO-like.

But the rarest of all the stratocumulus cloud species is the so-called 'roll cloud', stratocumulus volutus, which is closely related to the supplementary feature known as arcus, or shelf cloud. Unlike arcus, which is always attached to its parent storm cloud, volutus is free-floating, taking the form of a long, detached, tube-shaped cloud mass that often appears to roll slowly about a horizontal axis. Sc vol usually occur singularly but have occasionally been observed in parallel formations.

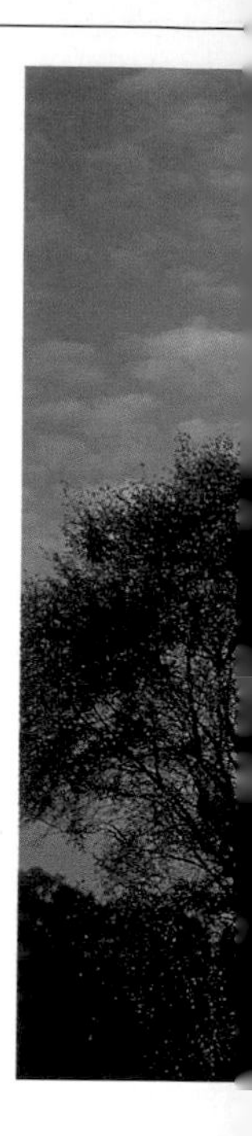

Modifications and transitions

The same cloud species can also appear in a less continuous form, with blue sky visible between the elements, where pockets of cool air are sinking, as in the photograph on page 32 of a sky filled with ragged white clumps of stratocumulus stratiformis. This formation bears a slight resemblance to altocumulus stratiformis (see C_M5), the cloudlets of which, however, are smaller and higher, as well as more clearly defined than these, so the two species ought to be easily distinguishable.

Weather implications

The layer clouds of stratocumulus stratiformis on page 32 look as if they might be threatening heavy rain, but in fact they would need to grow much thicker and darker in order to produce anything but the lightest scattering of drizzle.

Stratocumulus stratiformis, base 760m (2,500ft), in an evening sky over Anglesey, North Wales.

Patches of blue sky visible between the stratocumulus cloudlets show where pockets of cold air are beginning to sink.

C_L6

SPECIFICATION: Stratus in a more or less continuous sheet or layer, or in ragged sheets, or both, but no stratus fractus of bad weather.

Symbol = —

STRATUS NEBULOSUS

Appearance and nature

Stratus nebulosus is a low-level cloud formation made up entirely of water droplets, that appears as a uniformly grey, featureless layer that can extend across the sky for many miles. In contrast to convective cumuliform clouds, low stratus clouds form in cool,

stable conditions, with no pockets of turbulent activity going on. Instead, they tend to form when gently rising breezes carry cool, moisture-laden air across a cold sea or land surface, causing widespread, low-level condensation to occur, generally well below c. 500m (1,640ft). As can be seen in the photograph of Boston, Massachusetts, it is often low enough to obscure the tops of trees and buildings. Although these clouds can sometimes form from an overnight ground fog lifted by the breeze, it is worth bearing in mind, for the purposes of identification, that fog and mist are not strictly clouds at all, since they make direct contact with the ground.

Modifications and transitions

Watching the behaviour of stratus clouds over time can offer useful indications of short-term weather to come. If a layer of stratus forms in air that is lifting over hill slopes, it may well be followed by rain, something that people who live in valleys know well. When low stratus forms at night during the summer, the following morning can often start off gloomy, but the rising sun will evaporate the water droplets, seeing off the early cloud, and leaving the rest of the day fine and clear.

Weather implications

Stratus nebulosus comes in a variety of thicknesses, sometimes completely obscuring all light from above – when the designation would be stratus nebulosus opacus – and sometimes allowing the sun or moon to be seen with a clear

A layer of stratus nebulosus (St neb) draped over Boston, Massachusetts, obscuring the tops of buildings, and no doubt making downtown residents feel decidedly 'under the weather'.

outline, as can be seen in the atmospheric photograph below, of moonlight shining through a misty veil of stratus nebulosus translucidus. Unlike its distant cousins, watery altostratus and icy cirrostratus (see C_M1 and C_H5 to C_H6), low stratus clouds such as these do not produce optical phenomena such as haloes, parhelions and coronas, and neither are they capable of producing much in the way of rain.

Stratus clouds are often broken up by evaporation from a rising sun, or the arrival on the scene of a layer of warm, turbulent air. When in the process of breaking upwards, stratus clouds can sometimes appear more like low-lying

Moonlight shining through a thin veil of stratus nebulosus translucidus (St neb tr).

cumuliform clouds, as seen in the photograph above, of a thick layer of stratus nebulosus rising on convective currents, on their way to forming stratocumulus stratiformis clouds.

Stratus nebulosus breaking up at 275m (900ft) under the influence of rising air currents.

C_L7

SPECIFICATION: Stratus fractus of bad weather or cumulus fractus of bad weather, or both (see also pannus cloud, page 99).

Symbol = – – –

A ribbon of stratus fractus (St fra) wreaths the hills of Gorkhi Terelj National Park, Mongolia.

STRATUS FRACTUS

Appearance and nature

Stratus fractus are shreds of low cloud that can either appear separately, as in the photograph of a ribbon of stratus fractus wreathing the hills of Gorkhi Terelj National Park, Mongolia, or beneath the base of another, precipitating cloud layer, such as altostratus or nimbostratus (C_M2), when they are collectively referred to as 'pannus' (see page 99).

Wisps of stratus fractus cloud swirl in advance of oncoming rain.

Modifications and transitions

These ever-changing cloud fragments may merge and become a more or less continuous layer, sometimes even obscuring an extent of sky above. But usually they remain in the form of wisps of fast-moving cloud that breeze in just ahead of the rain, as in the photograph above. The incoming cloud is not fog or mist, as it has clearly formed well above ground, but has swept horizontally into the side of the intervening hill, over which it will quickly pass to make way for the imminent rain that it heralds.

Weather implications

Known also as 'scud' clouds, or 'messenger clouds', stratus fractus itself rarely produces rain, and at most is capable only of short-lived bouts of drizzle.

When stratus fractus is accompanying a rainy cloud cousin such as nimbostratus (C_M2), it can sometimes seem as if rain is falling from the lower stratus cloud, when in fact the rain is falling through it from above.

C_L8

CUMULUS AND STRATOCUMULUS AT DIFFERENT HEIGHTS

SPECIFICATION: Cumulus and stratocumulus other than that produced by the spreading out of cumulus (C_L4), with the bases of the two cloud types being at different levels.

Symbol = ⌓

Rising cumulus clouds thrust their way towards a layer of stratocumulus 300m (1,000ft) above them.

Appearance and nature

As described earlier, stratocumulus clouds can form in two distinct ways: either by the spreading out of the upper parts of cumulus congestus clouds, or from the breaking up of rising layers of stratus nebulosus by upward convection. The clouds in this classification of sky are both of the latter (convective) variety, being

small cumulus clouds forming beneath an already present layer of stratocumulus stratiformis, a relatively common situation that can clearly be seen in the photograph.

Modifications and transitions

As the rising cumulus clouds approach the level of the stratocumulus layer, they do not spread out horizontally, as is the case with stratocumulus cumulogenitus clouds (C_L4) but tend to thrust their way through the upper layer, as is beginning to happen in the photograph of stratocumulus at *c*. 900m (3,000ft) being approached by cumulus clouds *c*. 300m (1,000ft) below. Often the stratocumulus layer will thin out or break up around the point of contact with the ascending cumulus cloud.

Luke Howard described the changing face of this mixed-cloud sky in some detail in his 1803 essay *On the Modifications of Clouds*, observing how the upper cloud layer 'becomes denser and spreads, while the superior part of the cumulus extends itself and passes into it.' The result of such mixing, as the photograph shows, can be a complex and fascinating aerial landscape.

Weather implications

The lower cumulus clouds are unlikely to produce rain or drizzle, but the upper stratocumulus layer is worth keeping an eye on, as it could lift and thicken into a dark, gusty, rain bearing layer should moist air move in.

C_L9

CUMULONIMBUS CAPILLATUS

SPECIFICATION: **Cumulonimbus, the upper part of which is clearly fibrous (or cirriform), often in the shape of an anvil. This cloud may sometimes be accompanied by cumulonimbus calvus (without anvil) (see C_L3).**

Symbol = ⏅

Appearance and nature

With their towering thunderhead, cumulonimbus capillatus clouds can grow to vast heights, extending from their low bases, 600m (2,000ft) or lower, to 18km (60,000ft) or more above the ground – to where the troposphere meets the stratosphere – making them by far the tallest structures on Earth.

Cumulonimbus capillatus anvil rearing over Cape Town, South Africa.

No wonder the expression 'to be on cloud 9' means 'to be on top of the world'.

The true scale of this enormous cloud is best appreciated when viewed against a flat horizon, as in the dramatic photograph above of a vast single-cell cumulonimbus structure, complete with its icy striated incus (Latin for 'anvil'), rearing over a small farming settlement in Texas.

Cumulonimbus clouds are usually an energetic development of cumulus congestus clouds (see C_L2), which have grown unstable

With as much energy as ten Hiroshima-sized atom bombs, storm-clouds such as this can grow to vast heights, towering over the landscape below.

through powerful upward convection currents. As was seen in the entry for cumulonimbus calvus (C_L3), which represents the intermediate phase between cumulus congestus and the full-blown cumulonimbus capillatus seen here, the characteristic cauliflower-shaped summit of the congestus cloud has been replaced by the glaciated anvil, a vast canopy of ice crystals which can sometimes be spread for hundreds of kilometres by the strong winds of the upper atmosphere, creating an eerie mushroom-cloud effect, as can be seen in the photograph on page 46.

Modifications and transitions

Individual cumulonimbus clouds, of the kind shown here, are relatively short-lived single-cell varieties, lasting perhaps an hour or more before raining or blowing themselves out. If a number of single-cell cumulonimbus clouds manage to coalesce, however, they can form a multicell or even a supercell storm structure that can last for many hours, or even for days, and it is these great storm systems, with their extreme vorticity, that are responsible for producing the thousand or more tornadoes that batter the American Midwest every year.

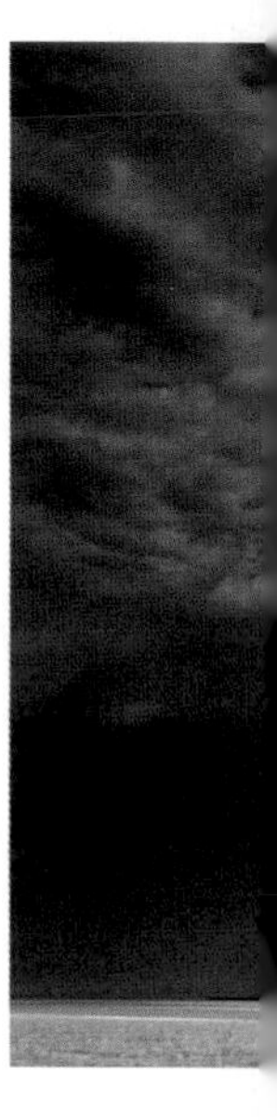

Weather implications

Thunder, lightning, hail, heavy rain and strong winds are the likely accompaniments to this vigorous cloud, which during its phase of maximum power can contain as much energy as ten Hiroshima-sized atom bombs.

From afar, cumulonimbus capillatus and cumulonimbus calvus are unmistakeable, but viewed from immediately below the cloud base, where the entire sky is likely to be dark and lowering, and busy with pannus clouds (see page 99), both can easily be confused with nimbostratus (C_M2), the thick, grey blanket associated with persistent rain or snow.

Unless it is actually hailing or thundery (the giveaway signs of a cumulonimbus cloud), it is the quality of the rain that will assist identification: widespread, persistent rain is most likely to fall from a nimbostratus cloud, while sudden, heavy bursts or sharp showers are more likely to come from a cumulonimbus. From immediately below a cumulonimbus cloud, of course, the top cannot be seen, making identification of the species (calvus or capillatus) impossible, unless you telephone someone in a nearby town who can see the cloud from the side … in that situation, the classification C_L=9 is used.

A large cumulonimbus capillatus cloud passes over farmland, its volume rapidly diminishing through heavy precipitation.

C_M1

SPECIFICATION: **Altostratus, the greater part of which is semi-transparent; through this part the sun or moon may be weakly visible, as through frosted glass.**

Symbol = ∠

ALTOSTRATUS TRANSLUCIDUS

Appearance and nature

Altostratus translucidus is a featureless layer of thin, grey-blue cloud that can spread to cover most of the sky, giving rise to dull, overcast conditions. It can form from a descending veil of cirrostratus (see C_H5), as well as, less commonly, from the spreading out of the upper parts of a cumulonimbus cloud. But it is usually the result of the lifting of a large mass of warm air ahead of an incoming warm front or occlusion.

Formed by the lifting of a large mass of warm air, altostratus translucidus clouds form a dull, featureless layer across the sky.

A sunset seen through a thin layer of altostratus translucidus (As tr).

Modifications and transitions

When the sky is filled with thin altostratus, sunlight becomes diffused and watery, and rarely casts a shadow on the ground. A rising or setting sun will glow pink or orange through the cloud veil, as seen in the image of a sunset above, while optical phenomena such as coronas and irisation are common accompaniments (see pages 129–130), due to the cloud being largely made up of uniformly sized water droplets.

Weather implications

If the warm front continues to advance, pushing further moist air upwards, altostratus translucidus can thicken into altostratus opacus or nimbostratus (see following entry), a sure sign of imminent rain.

C_M2

SPECIFICATION: Nimbostratus is a thickened, rainy form of altostratus opacus (also C_M2), the greater part of which is dense enough to hide the sun or moon from view

Symbol = ⁄⁄

ALTOSTRATUS OPACUS, OR NIMBOSTRATUS

Appearance and nature

Nimbostratus usually begins life as a high, thin layer of altostratus translucidus (see previous entry) which descends and thickens as it takes on rising moisture imported by an advancing warm front. As the cloud thickens, sometimes in several layers, the sun or moon will become increasingly obscured, and the altostratus translucidus variety will be redefined as an altostratus

opacus, in which weak sunlight struggles to appear behind the thickening veil of cloud. Even if the cloud layer grows so thick that the sun or moon is screened out entirely, the cloud is not technically a nimbostratus until rain starts to fall from it.

Nimbostratus is unique among the ten cloud genera for having no species or varieties to its name, although it can occur at a range of altitudes, despite the fact that it is classified as a medium-level cloud.

Modifications and transitions

Both altocumulus opacus and nimbostratus can be very long-lasting clouds, and over time the air below them can become highly saturated, which is when ragged shreds of pannus cloud develop beneath the main cloud layers.

Weather implications

With the exception of gardeners at the height of a hosepipe ban, it is a safe bet that nimbostratus would turn out to be everybody's least favourite kind of cloud. Grey, gloomy and pouring with rain, this cloud blanket seems to last all day (especially at weekends and at major sporting fixtures).

At its lowest, with near-continuous sheets of rain, snow or sleet serving to diffuse any remaining light, nimbostratus can appear to be connected directly with the sodden landscape below, leaving us feeling not just under the weather, but in it.

A ragged band of altostratus opacus, thickening into rainy nimbostratus.

C_M3

SPECIFICATION: **Altocumulus, the greater part of which is semi-transparent; the various elements of the cloud change only slowly and are all at a single level.**

Symbol = ω

ALTOCUMULUS TRANSLUCIDUS AND ALTOCUMULUS STRATIFORMIS AT A SINGLE LEVEL

Appearance and nature

Altocumulus clouds are a group of medium-level cloud species that can occur as rounded, individual 'bread-roll' clouds, either in localised patches or in vast, extensive layers, as well as in haunting 'UFO' formations (see altocumulus lenticularis, C_M4). They are usually composed of either supercooled water droplets or ice

Altocumulus stratiformis clouds arranged into parallel bands.

Altocumulus translucidus can form in rounded 'bread-roll' cloudlets, such as those shown on this page.

crystals, or a mixture of both, so they are equiped to exhibit a wide range of optical phenomena, depending on which form of water is predominant (see the entries on optical phenomena on pages 125–137).

Modifications and transitions

Altocumulus clouds are so varied in their arrangements that they occupy seven of the 27 categories of sky. Despite their variant formations – radial, fleecy and cellular – all clouds of C_M3 exhibit the same degree of translucency, and all can clearly be seen to persist as a single layer of cloud, in which the individual elements change very little over time.

Weather implications

These tend to be long-lasting formations that persist in stable weather conditions, so as long as there are no visible changes observed the weather will remain calm. However, any observed thickening of the cloud layer, giving a frosted glass appearance, would suggest that rain is likely to be on its way within the next 24 hours.

C_M4

SPECIFICATION: Patches of altocumulus, often lenticular ('lens- or almond-shaped'), the greater parts of which are semi-transparent; these clouds occur at more than one level, and are continually changing in appearance.

Symbol = ␣

ALTOCUMULUS LENTICULARIS

Appearance and nature

These elegant lenticular ('lens-shaped') altocumulus clouds are formed when a flowing layer of moist air is uplifted by the slope of an intervening hill or mountain. The wind that carries the air over a mountain rises gently, cooling unevenly as it does so, sending bouncing waves of moisture-laden air streaming away from the obstruction. Lenticular wave clouds form in the crests of these air waves.

UFO-shaped lenticular clouds in an eerie pile d'assiettes ('stack of plates') formation.

Modifications and transitions

These beautiful undulating clouds often emerge and dissipate in unexpected ways, according to the movement of the air currents, sometimes appearing UFO-shaped. Sometimes they appear in stacked layers, like the formation above, known as a *pile d'assiettes*, from the French for 'stack of plates'. They are always localised phenomena, however, and never go on to invade the sky like their near relatives altocumulus stratiformis (see following entry, C_M5).

Weather implications

Given that these clouds tend to be created orographically (that is, by the presence of mountains in stable airflows) their presence indicates stability in the local weather conditions. Should they suddenly disappear or begin to spread across the sky, a change in the weather can be expected over the next few hours.

A heart-shaped altocumulus lenticularis (Ac len) formed by a layer of moist air uplifted over some hills.

C_M5

ALTOCUMULUS STRATIFORMIS

SPECIFICATION: Semi-transparent altocumulus, either in bands, or in one or more fairly continuous layers, which progressively invade the sky, growing thicker as they do so.

Symbol = ῳ

Appearance and nature

Although similar in appearance to the clouds of C_M3 (altocumulus stratiformis at a single level) this is another example of how cloud classification takes transformation over time to be one of its identifying characteristics. Still photographs cannot really do justice to the distinction between these clouds and those of the C_M3 coding. C_M3 clouds are much more static and localised, while C_M5 progressively invade the sky, gathering overhead rapidly, and thickening as they do so, sometimes creating an entirely covered sky that stretches from horizon to horizon.

A dappled sky filled with thickening cloudlets of altocumulus stratiformis.

An evening sky filled with noticeably thickening altocumulus stratiformis cloudlets.

Modifications and transitions

These clouds often progess in parallel bands that spread through gentle turbulence into a variety of attractive formations. Their advancing edges may also consist of small cloudlets that thicken progressively. This thickening is often due not to the cloudlets themselves growing in size, but to the increased concentration of water droplets present within them, lending them an 'optical depth' that is sometimes enough to completely obscure the sun or moon.

Weather implications

These clouds are worth keeping an eye on, since they, or at least parts of them, are liable to thicken and descend, transforming into altostratus opacus clouds, or even nimbostratus, at which the coding would change to C_M7 and then, depending on what happens next, to a rain-sodden nimbostratus (C_M2).

C_M6

SPECIFICATION: **Altocumulus resulting from the spreading out of cumulus (or cumulonimbus) clouds.**

Symbol = ⋏

Altocumulus cumulogenitus (Ac cugen) in its early stage of formation, in which patchy cloudlets can be seen spreading from the top of a cumulus congestus cloud.

ALTOCUMULUS CUMULOGENITUS

Appearance and nature

As is also the case with stratocumulus cumulogenitus clouds (C_L4), the upward growth of cumuliform clouds can be halted by the presence of a temperature inversion, at which they begin to spread out horizontally instead, to form an entirely new species of cloud. An early stage of this process can be seen in the photograph below, in which patchy cloudlets of altocumulus can be seen spreading

out from the top of a cumulus congestus cloud. If the spreading continues, the patches will tend to thicken and join, creating an extensive wedge-shaped layer of cloud.

This species of altocumulus can also be formed from the elements left behind by a decayed cumulonimbus cloud, a formation for which the designation would more correctly be altocumulus cumulonimbogenitus (Ac cbgen), although the coding would still be C_M6.

Modifications and transitions

Although altocumulus cumulogenitus can sometimes appear in the shape of an anvil, it never has the striations or the icy sheen of a true cumulonimbus anvil, as seen in the summits of C_L9 itself (see page 46). Sometimes, the spreading of this cloud is only temporary, and upward growth is soon resumed, so that the altocumulus appears to one side of the new, higher cumulus cloud.

Weather implications

Like the clouds of C_L4, these clouds spread out at the base of a temperature inversion, should they thicken and spread to cover a large area of sky, they could produce some precipitation, although this would tend to be light and short-lived, since altocumulus clouds are not usually associated with heavy rain or snow.

Altocumulus cumulogenitus spreading laterally, in the shape of an anvil. The cloud's upward growth has been arrested by a temperature inversion in the upper atmosphere.

C_M7

SPECIFICATION: **Altocumulus translucidus, stratiformis or opacus in two or more layers, not progressively invading the sky; or a single layer of altocumulus opacus or altocumulus stratiformis, not progressively invading the sky; or altocumulus appearing with altostratus and/or nimbostratus.**

Symbol =

ALTOCUMULUS STRATIFORMIS DUPLICATUS, OR ALTOCUMULUS WITH ALTOSTRATUS OR NIMBOSTRATUS

The specification C_M7 is used to describe three closely related varieties of sky.

Altocumulus stratiformis duplicatus (Ac str du) in more than one layer, creating a richly patterned evening sky, with the upper layer lit from below, contrasting with the deeply shadowed bottom layer, 1.2km (4,000ft) below.

Appearance and nature

Altocumulus stratiformis duplicatus is when patches or sheets of altocumulus stratiformis appear in more than one layer, as shown in the photograph on page 62 of a dramatic evening sky, in which two distinct layers of altocumulus stratiformis (recognisable from the C_M5 category) have formed, one at 2.4km (8,000ft), the other at 3.6km (12,000ft).

Altocumulus stratiformis also occurs in a single layer, an example of which can be seen in the photograph on page 63.

Altocumulus in tandem with altostratus, or even nimbostratus, as seen in the photograph on page 63, results from local transformation processes, through which altocumulus clouds acquire the appearance of altostratus.

Altocumulus with altostratus.

Modifications and transitions

Altocumulus stratiformis duplicatus clouds do not change continually, nor do they invade the sky.

Thick, single-layer altocumulus stratiformis often occur in localised patches, and do not progressively invade the sky; their cloudlets can bear a superficial resemblance to those of cirrocumulus floccus (C_H9), although altocumulus elements will tend to exhibit three-dimensional shading; any doubt can usually be dispelled by holding up one's hand at arm's length, and measuring the width of a cloudlet with one's fingers: those of altocumulus will typically measure two or three fingers' wide, while those of cirrocumulus will usually only measure one.

Altocumulus with altostratus or nimbostratus can appear in two or more layers, with each layer showing certain characteristics of both species, as in the Isle of Skye image opposite, in which a more stratiform layer rides over a slightly more cumuliform layer.

A localised patch of thick altocumulus stratiformis (Ac str).

Altocumulus with altostratus, in distinct layers, their bases between 3 and 4.5km (10,000 and 15,000ft), Isle of Skye from Mallaig, Scottish Highlands.

Weather implications

The combination of cloud layers in altocumulus stratiformis duplicatus, even if they are individually quite thin and diffuse, can sometimes prove dense enough to mask the sun or moon completely, and they can be quite persistent, lasting until sunset.

ALTOCUMULUS CASTELLANUS, ALTOCUMULUS FLOCCUS, ALTOCUMULUS VOLUTUS

SPECIFICATION: Altocumulus with sproutings either in the form of towers or castellations (castellanus) or small cumuliform tufts (floccus). Also roll cloud (volutus).

Symbol = M

Appearance and nature

The cloudlets that make up altocumulus castellanus formations exhibit sproutings in the form of small towers or battlements, a sure sign of instability in an upper layer of the sky. The cloud elements themselves have a common base and sometimes appear to be arranged in lines, as can be seen in the photograph, on the left, of a wind-ordered file of these clouds marching against an early evening sky.

A wind-ordered file of altocumulus castellanus (Ac cas) emerges from a layer of altocumulus floccus (Ac flo) clouds over the northern suburbs of Mumbai.

Altocumulus floccus (which can sometimes form from the dissipated bases of altocumulus castellanus) appear as white or grey scattered tufts with rounded and slightly bulging upper parts, resembling small ragged cumulus clouds. They often feature fibrous trails of virga (rain or snow not reaching the ground, see page 104) trailing from their bases.

The rarest of all the altocumulus clouds is volutus (or 'roll-cloud'), which takes the form of a detached, tube-shaped cloud mass that often appears to roll slowly about a horizontal axis. Caused by differences in wind speed and direction between the base of the cloud and its summit, Ac vol usually occurs as a single line and seldom extends from horizon to horizon.

Evening displays of altocumulus floccus will often precede wet or stormy weather in the morning.

Modifications and transitions

Altocumulus castellanus can also descend and merge to form large cumulus (C_L2) or sometimes even cumulonimbus clouds (C_L3 or C_L9).

Weather implications

The larger the castellations, the more vigorous the instability, and a sighting of altocumulus castellanus clouds can be a reliable indication of thunderstorms coming in over the next 24 hours.

The floccus species of altocumulus is also associated with humid, unstable conditions, likely to lead to thundery conditions developing over a wide area (as opposed to local thunderstorms originating from cumulonimbus clouds immediately overhead). An evening display of them will often precede wet or stormy weather in the morning, especially if convection kicks in with the rising sun, and young, energetic cumulus clouds end up joining in with the moisture-laden altocumulus clouds already present.

Tufts of altocumulus floccus with virga (Ac flo vir) – fallstreaks of icy rain or snow which evaporate before reaching the ground.

C_M9

SPECIFICATION: Altocumulus of a chaotic sky, generally at several levels.

Symbol = ◌

ALTOCUMULUS OF A CHAOTIC SKY

Appearance and nature

Altocumulus of a chaotic sky usually occurs at several levels. The sky is characterised by a heavy appearance, with broken sheets of poorly defined clouds at several transitional stages, from medium-level clouds or low, thick altocumulus, to high, thin altostratus.

Swirling clouds in a chaotic sky.

Clouds of many varieties appearing at several levels.

Modifications and transitions

In the first of the two photographs of altocumulus of a chaotic sky, a variety of cloud forms swirl around together. A turret of altocumulus castellanus even makes an appearance towards the right of the picture.

In the second image above, clouds of many varieties appear at several levels.

Weather implications

The weather prognosis is uncertain when the sky looks like this, although anyone seeing it would be well advised to keep an umbrella handy

C_H1

SPECIFICATION: Cirrus clouds in the form of filaments, strands or hooks, not progressively invading the sky.

Symbol = ⟶

CIRRUS UNCINUS & CIRRUS FIBRATUS

Appearance and nature

The high, white, delicate cirrus clouds of species C_H1 occur generally in curved filaments or straight lines, 'pencilled, as it were, on the sky', as Luke Howard described their appearance in 1803. Like all cirrus clouds, they are composed entirely of gently falling ice crystals, at altitudes above *c.* 6km (20,000ft).

Composed entirely of gently falling ice crystals, the curved filaments of cirrus uncinus (Ci unc) form easily identified commas in the sky.

Cirrus fibratus (Ci fib) clouds arranged into bands by the winds of the upper atmosphere.

Cirrus fibratus can often be arranged by the wind into parallel bands, as can be seen in the photograph above, which appear to converge towards the horizon. Even in such relatively dense arrangements, cirriform clouds still appear wispy and diffuse compared to the majority of low and medium cloud formations, because the concentrations of ice crystals in cirrus clouds are so much lower than the droplet concentrations found in clouds that are composed of liquid water.

Modifications and transitions

Cirrus sometimes form from the virga (see page 104) of high cirrocumulus clouds, but are usually formed when layers of relatively dry air ascend in the upper troposphere, the small amount of vapour then subliming into ice when it meets its subzero dew point (sublimation is the process of transforming directly from a solid to a gas, or vice versa, with no intermediate liquid state). These smaller kinds of lofty cirrus cloud usually appear on their own in a dry blue sky, since if the air were more humid, other types of cloud would have formed at lower levels.

The mariner's wind-warning: 'Mares' tails' of cirrus uncinus being drawn into long filaments by the wind. The longer the filaments grow, the stronger the wind will become.

Delicate, lofty cirrus uncinus trace beautiful patterns in a dry, blue sky

Weather implications

If the cirrus does not begin to spread then fine weather may well continue for a while, but if the cirrus begins to increase its cover, thickening or spreading out horizontally, then it means that a warm front is on its way, pushing up moist air ahead of itself, and causing the weather to take an imminent turn for the worse. Sailors have long viewed the growth of comma-shaped cirrus clouds as a useful 'wind warning', and in the photograph on the left, the crystals falling from a group of 'mares' tales' (a popular name for cirrus uncinus) can be seen being drawn into long filaments by the wind: 'mare's tails and mackerel scales make tall ships carry low sails', as the old weather-motto had it.

C_H2

SPECIFICATION: **Dense cirrus, in patches or sheaves, which usually do not increase (spissatus); or cirrus with sproutings either in the form of small turrets or battlements (castellanus) or small cumuliform tufts (floccus).**

Symbol =

CIRRUS SPISSATUS, CIRRUS CASTELLANUS & CIRRUS FLOCCUS

Appearance and nature

Cirrus spissatus is a thick, dense species of cirrus cloud, which can often appear to dominate large areas of sky. The castellanus and floccus species that are also classed as C_H2 exhibit turretty sproutings or ragged patches, and often feature trailing filaments

Icy cirrus spissatus (Ci spi) clouds dominate a large area of sky.

below the main cloud, as can be seen in the photograph of cirrus castellanus cloudlets on the previous page, with what look like lengthy striated kite-tails hanging below. These tendrils form when the descending ice crystals end up in a deep layer of cold air that is moving at a stable speed, causing them to be spread across the sky, sometimes for enormous distances.

Modifications and transitions

Cirrus clouds can also form from the spreading out of aircraft contrails, which grow by seeding moisture present in the air, and which can sometimes last for many hours, becoming indistinguishable from naturally occurring clouds (see page 120).

Weather implications

Thickened or actively changing cirrus spissatus will often appear in the vanguard of an approaching warm front, as a parcel of moist air is forced up over the shallow wedge of cooler air that it encounters, causing high clouds to form in the upper atmosphere. It often means that wet or stormy weather will be on its way some time over the following 48 hours.

Although cirrus castellanus and cirrus floccus spend their careers slowly falling, and are thus technically 'precipitating clouds', they rarely produce any kind of precipitation that reaches the ground, although they often exhibit virga (see page 104), trails of descending snow or rain that evaporate in the warm lower air, long before they are able to reach the surface.

In this photograph taken in Devoran, Cornwall, icy cirrus spissatus (Ci spi) clouds grow to dominate a large area of sky.

C_H3

SPECIFICATION: Dense cirrus, often in the form of an anvil, being the remains of the upper parts of a cumulonimbus cloud.

Symbol = ⌐

CIRRUS SPISSATUS CUMULONIMBOGENITUS

Appearance and nature

This form of cirrus derives from the leftover anvil of a decayed or rained-out cumulonimbus capillatus cloud (C_L9) (see also incus, page 102). The cumulonimbus anvil is essentially a vast canopy of turbulent cirrus that

can sometimes be carried away from the main stormcloud by the high-speed winds of the upper atmsophere, as has clearly happened in the photograph of a marooned anvil flailing above Reading, in Berkshire. The resultant C_H3 cirrus clouds are usually frayed and battered at the edges, but are still sufficiently thick to veil the sun, and therefore grey in colour, in contrast to the characteristic delicate whiteness of cirrus uncinus (C_H1) or cirrus castellanus (C_H2) clouds.

Modifications and transitions

Other cirriform clouds may also be present at the same time as C_H3, so it is sometimes hard to be absolutely certain that the cloud is truly cumulogenitus, rather than merely an energetic example of cirrus spissatus (see previous entry). In uncertain situations such as this – especially if one has not actually witnessed the transformation of the anvil at first hand – it is perfectly acceptable to use the simpler cirrus spissatus (C_H2) designation.

Weather implications

Because it is not always easy to know whether a particular form of cirrus spissatus originated from a rained-out storm cloud, or is simply a thick, energetic cirrus variety, its weather prognosis is hard to read. However, any kind of cirrus cloud that thickens and spreads over the course of the day is a reliable indicator of bad weather to come within the next 24–48 hours.

The leftover anvil of a rained-out cumulonimbus cloud, cirrus spissatus cumulonimbogenitus (Ci spi cbgen) is nevertheless a cloud species in its own right.

C_H4

SPECIFICATION: Cirrus in the form of hooks (uncinus) or filaments (fibratus) (see clouds of C_H1) progressively invading the sky, generally thickening as they do so.

Symbol = 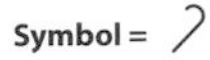

CIRRUS UNCINUS OR CIRRUS FIBRATUS PROGRESSIVELY INVADING THE SKY

Appearance and nature

The cirrus clouds of C_H4 are the same species as those of C_H1, but with the added feature of their progressively invading the sky, due to the influence of an approaching warm front as it serves to slide expanses of warm moist air over wedge-shaped areas of colder air, causing icy clouds to form at high altitudes (around 6km/20,000ft).

Modifications and transitions

These clouds often seem to fuse together in the direction from which they first appear, as in the photograph on the previous page of spreading cirrus fibratus, but they can also arrange themselves into great rippling parallel formations, as in the stunning example of invading cirrus on page 84. They will often thin out over time to form uniform veils of cirrostratus (see following entries).

Dense formations of cirrus uncinus fill the evening sky, promising a bout of bad weather in the morning.

'Mares' tails and mackerel scales make tall ships carry low sails': these characteristic mares' tails of cirrus uncinus (Ci unc) invading the sky are a sure sign of turbulent weather to come.

Bad weather on its way: cirrus fibratus (Ci fib) invading the sky, in advance of an approaching depression.

Weather implications

When cirrus clouds are seen to thicken and increase, sometimes spreading to entirely cover the sky, it is a sure indication of an advancing depression, an area of low pressure and rising air that is usually associated with imminent bad weather. Ice crystal-related optical phenomena such as haloes, mock suns and circumzenithal arcs (see pages 125–128) are occasional accompaniments to this attractive species of sky, and serve as further indications of the deteriorating weather conditions to come.

Cirrus fibratus clouds can sometimes arrange themselves into parallel bands, as in this example.

C_H5

The leading edges of these bands of cirrostratus (Cs) can clearly be seen as they invade the sky below 45°.

SPECIFICATION: **Cirrus, often in bands, with cirrostratus, or cirrostratus on its own, progressively invading the sky, and growing denser as it does so, but the continuous veil of cloud does not reach 45° above the horizon.**

Symbol = ⟂

CIRROSTRATUS (BELOW 45°)

Appearance and nature

Cirrostratus clouds are high ice-crystal clouds produced by the slow ascent of air in the troposphere. Like cirrus clouds, from which they often evolve, they are formed when water vapour sublimes into ice crystals, often ahead of advancing weather fronts, at altitudes above 6km (>20,000ft).

Skies of C_H5 will often feature bands or tufts of thickened cirrus appearing ahead of the cirrostratus, but the principal component is always an advancing veil of whitish cirrostratus slowly advancing over the horizon, but (at least in this classification) no more than 45° above it.

Modifications and transitions

Cirrostratus formations of C_H5 and C_H6 sometimes have a definite 'edge' to them, as can be seen in the photograph, although they can also be fringed with cirrus clouds, making initial identification more difficult.

Weather implications

The movements of cirrostratus clouds are always worth keeping an eye on, because their behaviour can often give advance notice of changes in the weather to come. If it is seen to form out of spreading cirrus (C_H4) that grows thicker and more continuous, it is likely to be followed by wet weather within 48 hours, but if gaps begin to appear in the cirrostratus veil, and it begins slowly to change into cirrocumulus (C_H9), the weather will probably continue to be dry for a day or so (but remember to keep an eye on the developing cirrocumulus).

C_H6

SPECIFICATION: Cirrus, often in bands, with cirrostratus, or cirrostratus on its own, progressively invading the sky, and growing denser as it does so, with the continuous veil of cloud extending more than 45° above the horizon, although without the sky being totally covered.

Symbol = 2

CIRROSTRATUS (ABOVE 45°)

Appearance and nature

If the cirrostratus of C_H5 advances further, becoming continuous to more than 45° above the horizon, but without covering the entire sky, its appearance is coded as C_H6 (although the distinction between C_H5 and C_H6 may seem a little academic for the purposes of everyday cloud identification). But the cloud itself changes as it advances, tending to grow denser as it proceeds, revealing a more fibrous structure, and with a less clear-cut leading edge, as can be seen in the photographs.

Modifications and transitions

As with the sky of C_H5, it is often accompanied by neighbouring cirrus clouds, examples of which can easily be seen in the main photograph opposite.

Cirrostratus, (Cs) invading the sky and above 45° elevation, its leading edge now less clear-cut than in the preceding classification.

Weather implications

As with the previous entry, the behaviour of these clouds is worth noting carefully, as cirrostratus clouds tend to form ahead of advancing weather fronts. Though these thin veils of icy cloud do not produce rain themselves, they can give way to thicker forms of cirrus cloud, the spreading of which is a classic sign of bad weather to come.

Cirrus clouds accompany this spreading band of cirrostratus, as it progressively invades the sky.

C_H7

SPECIFICATION: **A veil of cirrostratus which covers the entire sky.**

Symbol = ⌣c

A dense layer of cirrostratus fibratus (Cs fib) that has advanced to cover much of the evening sky.

CIRROSTRATUS FIBRATUS OR CIRROSTRATUS NEBULOSUS INVADING THE WHOLE SKY

Appearance and nature

The C_H7 sky is entirely covered by a high-riding veil of cirrostratus, which can sometimes be quite dense and fibrous in appearance (cirrostratus fibratus) and at other times so thin that it goes completely unnoticed, having little or no direct effect on sunlight. This is cirrostratus nebulosus, and is shown on page 91, in the image of sunlight being gently diffused through a dappled gauze of cloud.

A thin veil of cirrostratus nebulosus (Cs neb) through which dappled sunlight can clearly be seen.

Modifications and transitions

At times the thinner species of cirrostratus can appear so milky and indistinct that the only clue that they are there at all is the sight of a halo around the sun or moon, an effect caused by light refracting through the clouds' hexagonal ice crystals (see page 125 for a more detailed description of halo phenomena).

Weather implications

This kind of cirrostratus is prone to thicken and descend at the onset of a warm front, transforming by degrees into altostratus (C_M1), which is often a precursor of nimbostratus (C_M2), the soggy culmination of the complex, day-long layer-cloud cycle that so often begins with the first appearance of the high, delicate mares' tails of cirrus uncinus (C_H1).

C_H8

CIRROSTRATUS NOT PROGRESSIVELY INVADING THE SKY

SPECIFICATION: A veil of cirrostratus which neither covers the entire sky, nor progressively invades it.

Symbol = ⌐

Appearance and nature

The sky state categorized as C_H8 features a veil of cirrostratus that is not, or is no longer, invading the sky progressively, and which does not cover it entirely, as distinct from the invasive categories of cirrostratus outlined earlier (C_H5 to C_H7). It typically appears as a broken patch of cirrostratus fibratus, as in the photograph below, with edges that are either clear-cut or fairly ragged in appearance.

Patchy cirrostratus clouds neither invading the sky nor entirely covering it.

Cirrostratus, in all its manifestations, has often lent its services to the creation of colourful sunsets, as can be seen in this dramatic example.

Modifications and transitions

In this kind of sky cirrus or cirrocumulus might also be present, without predominating over the main body of cirrostratus.

Weather implications

Any non-changing sky state indicates that prevailing weather conditions are stable. This variety of cirrostratus tends to be fairly long-lived but, as always, it is change that lets us know when things are beginning to happen. So it may well be that, after several hours have elapsed, the cloud veil will begin to spread in advance of a change in the weather (see C_H5 and C_H6).

C_H9

SPECIFICATION: **Cirrocumulus appearing alone, or with cirrus and/or cirrostratus, as long as the cirrocumulus is predominant.**

Symbol = ![symbol]

CIRROCUMULUS STRATIFORMIS, CIRROCUMULUS FLOCCUS & CIRROCUMULUS LENTICULARIS

Appearance and nature

Cirrocumulus clouds are a group of relatively rare cloud species, made up of ice crystals and supercooled water droplets which occur either in small, rippled, 'grainy' white patches or in shallow, extensive formations across the sky. They form high in the atmosphere, between *c.* 5 to 14km (16,000 and 45,000ft) when turbulent upward air currents encounter high cirrus or cirrostratus clouds, transforming some of their ice crystals into supercooled water droplets, while breaking them up into the rounder cloudlets of cirrocumulus. This vestigial resemblance to cirrus is evident in the photographs on page 95, of a patch of 'herringbone' cirrocumulus stratiformis seen high in the sky.

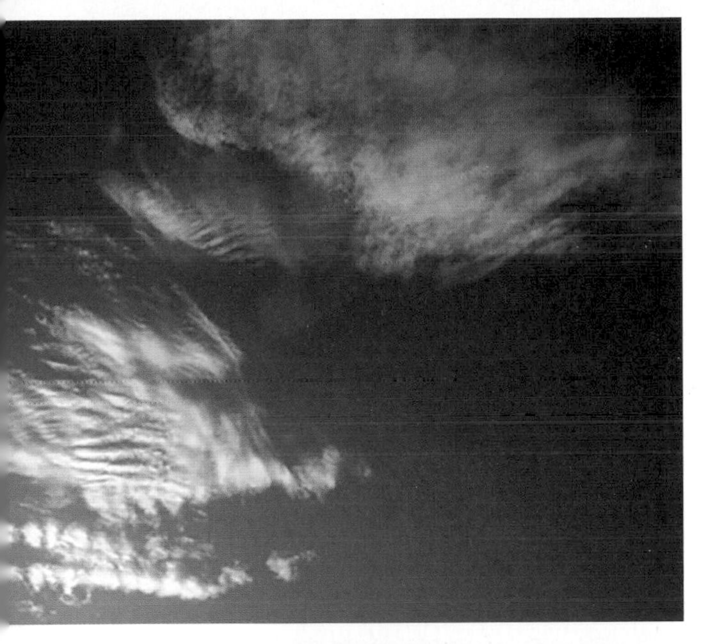

Small patches of 'herringbone' cirrocumulus stratiformis (Cc str), made up of ice crystals and supercooled water droplets.

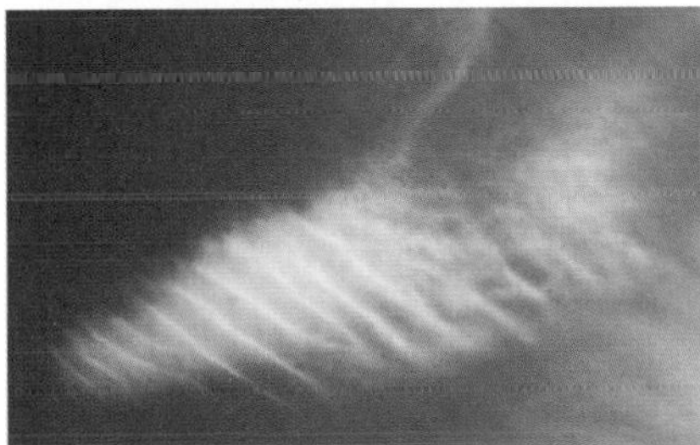

Cirrocumulus floccus form when the convection that gives rise to the cloudlets is stronger and more unstable, serving to emphasise the cumuliform structure of the new cloud formation, despite its cirriform origins.

In hilly areas, by contrast, the effects of orographic wave motion through a moist, stable layer of air high in the sky can produce the rare cirrocumulus lenticularis formation – a cloud not to be confused with the lower-lying, and slightly more common altocumulus lenticularis (C_M4).

Modifications and transitions

Due to the generally unstable conditions in which they form, these smaller appearances of cirrocumulus tend to be short-lived, either thinning and spreading out to form veils of cirrostratus, or joining up with others to cover large areas of sky, as can be seen in the photograph on page 97. It is sometimes difficult to distinguish these larger outbreaks of cirrocumulus

Strong upward convection is responsible for the fleecy appearance of these cirrocumulus floccus (Cc flo) cloudlets.

Cirrocumulus lenticularis (Cc len).

from their lower altocumulus cousins, although it should be noted that the cloudlets of cirrocumulus tend to look much smaller than those of altocumulus stratiformis (C_M5), since they are so much further away, while their contours are never shaded in the way that the thicker elements of altocumulus formations usually are. If in doubt, one easy test is to hold up a hand at arm's length and spread it against the sky – cirrocumulus cloudlets are rarely more than one finger-width across, whereas altocumulus cloudlets tend to be two or even three times that size.

Weather implications

Broad, rippled displays of cirrocumulus stratiformis are often referred to as 'mackerel skies', due to the cloudlets' resemblance to fish scales. Mackerel skies have long been known as harbingers of bad weather, especially by seafarers, whose ancient saying: 'mare's tails and mackerel scales make tall ships carry low sails', is a testament to the fact that a large amount of moisture borne so high in a cold sky is a visible indication of an advancing depression, while the broken-up quality of the cloud layer itself shows that conditions up there are turbulent.

Accessory clouds

PILEUS (PIL)

Cumulus congestus with pileus (Cu con pil), a frozen cap of moist air lying flat on top of the cloud.

From the Latin for 'cap', pileus is a supplementary layer of flattened cloud that sometimes appears above a cumulus congestus (C_L2) or cumulonimbus calvus (C_L3) cloud. It is formed by the rapid condensation of a layer of moist air that is pushed up over the cloud's main summit, where it freezes into a layer of icy fog. Pileus clouds are often short-lived, with the main cloud beneath them rising through convection to absorb them. A pileus or cap cloud should not be confused with an incus ('anvil'), the yet more visibly hairy and striated area of ice crystals that develops above a cumulonombus capillatus cloud (C_L9).

PANNUS (PAN)

From the Latin for 'cloth' or 'rag', pannus are accessory clouds in the form of dark, ragged shreds of either stratus fractus (C_L7) or cumulus fractus (C_L1), that appear below other clouds, typically those of the genera altostratus, nimbostratus, cumulus and cumulonimbus. Pannus clouds have historically been referred to as 'scud' or 'messenger' clouds, particularly by sailors and farmers, for whom the message that they brought was: rain.

Pannus clouds are often attached to the main body of other, precipitating cloud layers, but they can sometimes form a distinct layer in their own right, obscuring the main cloud above them, as in this photograph, taken on a rainy day, of dense, dark pannus clouds swirling below a whitish layer of altostratus.

Stratus fractus pannus (St fra pan) cloud lowering above the town of Brynmawr, Gwent.

VELUM (VEL)

From the Latin for 'sail' or 'awning', a velum or veil cloud is a thin, wide, long-lasting layer of low altitude cloud through which the summits of cumulus congestus (C_L2) or cumulonimbus clouds will often pierce. In contrast to icy pileus clouds (page 98), which are generally short-lived, velum clouds are formed in layers of stable, humid air that are lifted by the convection currents within the main cumuliform clouds, and they can persist even after their host clouds have dispersed or decayed.

A veil cloud (velum) pierced by a rising cumulus congestus cloud.

FLUMEN (FLM)

Flumen ('flow cloud') are bands of low clouds associated with severe convective cumulonimbus storm clouds, arranged parallel to the low-level winds and moving into or towards the storm cell. The cloud elements are drawn towards the updraft into the supercell, the base being at about the same height as the updraft base. Note that flumen are not attached to the murus wall cloud (see murus, page 108) and that the cloud base is higher than the wall cloud. There is a variety of inflow band cloud known as the 'Beaver's tail', which is distinguished by its broad, paddle-like appearance, as can be seen in the accompanying photograph.

Flumen appearing near the base of a cumulonimbus storm cloud, with its distinctive 'beaver's tail' formation.

Supplementary features

INCUS (INC)

The anvil-shaped summit of a large cumulonimbus capillatus cloud (C_L9), the incus is an icy canopy that can grow to enormous heights above the main body of the cloud, spreading out laterally when it meets the tropopause – the atmospheric boundary separating the troposphere from the stratosphere – to create the characteristic flattened thunderhead seen here. It can be smooth in appearance, especially at a distance, but it is usually highly fibrous and striated, being composed of billions of ice crystals borne aloft by vigorous upward convection.

The incus can sometimes precede the main storm cloud by many miles, even producing cloud-to-ground lightning in the apparent absence of the main storm cloud. If, as sometimes happens, the main cumulonimbus cloud decays or rains itself out, leaving the anvil behind on its own, the resulting cloud is known as cirrus spissatus cumulonimbogenitus (see C_H3, page 80).

The flattened, icy thunderhead of the cumulonimbus capillatus (Cb cap inc).

MAMMA (MAM)

Mamma (also known as 'mammatus') are udder-like protuberances that can form on the under surfaces of stratocumulus or cumulonimbus clouds, particularly underneath the anvils of the latter. They are caused by powerful downdraughts, when pockets of cold, moist air sink rapidly from the upper to the lower parts of the cloud, reversing the usual cloud-forming pattern of the upward convection of warm, humid air. Their shapes and forms can vary considerably, from near-spherical pouches to tubular, rippled or merely undulating globules, arranged in cellular formation.

This dramatic image, taken at Brize Norton airfield in Oxfordshire, shows large mamma formed beneath a powerful cumulonimbus capillatus cloud (see C_L9, page 46).

Undulating mamma forming below cumulonimbus capillatus cloud.

VIRGA ('FALLSTREAKS') (VIR)

Virga (from the Latin for 'rod') is any form of precipitation, whether rain, snow or ice, that evaporates before it reaches the ground. Its failure to carry on all the way down is usually due to its passing through a layer of warmer or dryer air, although sometimes atmospheric conditions will change, and the virga will be replaced by real precipitation from the same cloud (see page 106).

Often wispy or hooked in appearance, virga is most associated with high- or medium-level cloud formations, as in this photograph of tendrils of virga falling from an altocumulus floccus (C_M8).

Tendrils of virga falling from an altocumulus floccus cloud. Such fallstreaks will usually evaporate long before reaching the ground.

ARCUS (ARC)

A distinct shelf or roll of low cloud that can appear below a powerful cumulonimbus cloud, an arcus (from the Latin for 'arch') is formed by strong downdraughts of cold air, which spread out ahead of the oncoming stormcloud, pushing up layers of warm air nearer the ground. These form dense, horizontal rolls of cloud, some of which (as the photograph shows) can look quite eerie and menacing.

Cumulonimbus capillatus with arcus (Cb cap arc).

PRAECIPITATIO (PRA)

From the Latin for 'fall', the term *praecipitatio* is applied by meteorologists to a cloud from which any kind of rain, snow or hail manages to reach the ground, as distinct from virga (see page 104). Although the two examples shown here are from cumulonimbus clouds, precipitation can fall from a variety of other cloud types, including stratus, stratocumulus, altostratus and nimbostratus clouds.

A heavy shower falling from a cumulonimbus cloud.

A hail shower falling over Kansas. Hail forms when warm updraughts of air hurl descending ice pellets back up into the colder regions of the cloud. The pellets grow through collision and freezing, creating ever bigger stones.

MURUS (MUR)

Murus ('wall cloud') is another supplementary feature associated with strong cumulonimbus storm clouds. These features (which are sometimes known as 'pedestal clouds') are isolated lowerings attached to the storm cloud's rain-free base, and indicate an area of strong updraft in which rain-cooled air is pulled towards the storm cloud's core. Murus clouds thus mark the area of strongest updraft within the storm, and are characterised by strong winds; in fact, tornadoes and tuba ('funnel clouds') often form within wall cloud structures. Murus is also sometimes accompanied by a cauda (see next entry).

CAUDA (CAU)

Cauda ('tail') is a horizontal, tail-shaped cloud (not a funnel, see page 113) that can extend from the main precipitation region of a supercell cumulonimbus; it is typically attached to the wall cloud (see murus), with cloud motion moving away from the precipitation area and towards the murus to which it is attached. Most movement is horizontal, but some rising motion is often apparent as well. Ragged-looking cauda is easily mistaken for the accessory cloud flumen (see page 101): both are types of inflow bands, but the cauda feature is attached to the storm's wall cloud (murus), while the flumen feature can be significantly larger and appear higher up in the storm structure, where it feeds into the storm cloud's updraft.

Murus ('wall cloud'), accompanied by a distinctive cauda ('tail') at the base of a large cumulonimbus storm cloud.

CAVUM (CAV)

Cavum (otherwise known as a 'fallstreak hole' or 'hole-punch cloud') is an effect caused by the sudden freezing of an isolated patch of supercooled cloud, which falls away to leave a visible gap in its place. The result is usually a well-defined circular (though sometimes linear) hole in the high cloud layer, from which virga or wisps of cirrus can be seen to descend. Cavum is typically a circular feature when viewed from directly beneath, but may appear oval shaped when viewed from a distance. The physical cause of the phenomenon is not yet fully understood, though it seems to happen only in supercooled clouds, in which water droplets remain in liquid form even in subzero temperatures. It is likely that particulates from aircraft exhaust play a part in the creation of cavum, since supercooling often occurs when there are not enough freezing nuclei available for airborne water droplets to turn into ice. Cavum usually occurs in altocumulus and cirrocumulus clouds, though has occasionally been observed in stratocumulus.

Cavum, also known as a 'fallstreak hole', appearing in a layer of altocumulus stratiformis.

ASPERITAS (ASP)

Asperitas (from the Latin for 'roughened) are well-defined, wave-like structures that appear on the undersides of stratocumulus and altocumulus clouds. They are more chaotic and show less horizontal organisation than the variety undulatus, with localised waves that appear either smooth or dappled with smaller features, sometimes descending into sharp points, as if viewing a patch of roughened sea from below. Asperitas was first identified by members of the Cloud Appreciation Society, and was the first of the twelve new official cloud terms to be adopted by the World Meteorological Organization in 2017.

A swirling display of altocumulus stratiformis asperitas over the city of Tallin, Estonia.

FLUCTUS (ALSO KNOWN AS KELVIN–HELMHOLTZ WAVES) (FLU)

Fluctus (from the Latin for 'wave') is a relatively short-lived formation that occurs when the boundary between a warm air mass and a layer of colder air beneath it is disturbed by strong horizontal winds, causing the upper layer to move faster than the lower. This aerial turbulence, or wind shear, causes the 'crests' of the waves to move ahead of the main body of the cloud, leading to the characteristic wave formation seen here.

This rare, short-lived phenomenon was originally named after two nineteenth-century scientists who pioneered the study of turbulent flow: the Belfast-born physicist William Thomson, 1st Baron Kelvin (1824–1907), and the German physicist Hermann von Helmholtz (1821–94). The new name, 'fluctus', was added to the official cloud classification in 2017.

Fluctus (Kelvin–Helmholtz waves) form in distinct layers of turbulent air moving at different speeds.

TUBA ('FUNNEL CLOUD') (TUB)

A tuba (or 'funnel cloud') sometimes forms at the base of a cumulonimbus cloud when a column of swirling air begins to rotate, condensing ambient moisture into water droplets. This vortex begins to move downwards, creating a tapered cone or funnel shape that protrudes some distance below the cloud, although it is rarely strong enough to make contact with the ground. When it does, it usually takes the form of a weak landspout or waterspout, rather than a fully fledged tornado, which tends to develop from the large-scale rotation of a tropical supercell thunderstorm, in contrast to the weak vorticity of the cold air funnel cloud.

Funnel cloud.

Special clouds

NOCTILUCENT CLOUDS

Noctilucent clouds (NLCs), also known as polar mesospheric clouds, are the highest clouds in the Earth's atmosphere, occurring in the mesosphere (the layer immediately above the stratosphere) at altitudes above 80km (50 miles), at least four times higher than those of any other cloud in this book (with the single exception of stratospheric nacreous clouds).

NLCs are extremely rare, although they seem to have become rather less rare over the past 20 years, which may or may not be due to human activity. The best chance of spotting one is on a clear midsummer night somewhere between 50° and 65° latitude (north or south), such as northern Scotland or Scandinavia, when they are underlit by the sun from over the Earth's shoulder. Their name derives from the Latin for 'night-shining'.

Appearing as thin, milky-blue or silvery waves high in the sky, on the fringes of space, NLCs look as mysterious as they in fact are: they remain the least understood clouds of all, the mechanics of their formation in such dry, clear, intensely cold atmospheric conditions (-125°C/-193°F) having not yet been discovered, although many hypotheses have been advanced, including the idea that they seed themselves from meteorite debris, from dust blasted high into the stratosphere by major volcanic eruptions on Earth, or even from the constituent elements of space shuttle exhaust fumes. The news from the solar-powered spacecraft Mars Express in 2006, that clouds of CO_2 with a similar appearance to NLCs are to be found high in the atmosphere above the Red Planet, can only add to the ongoing noctilucent mystery.

The least understood cloud of all, noctilucent clouds appear as thin, milky waves of colour high in the sky on midsummer nights.

NACREOUS ('MOTHER OF PEARL') CLOUD

Nacreous clouds, also known as polar stratospheric clouds (PSCs), appear high in the atmosphere, some 15 to 30km (10 to 20 miles) above the Earth, generally in latitudes higher than 50°, particularly in the northern hemisphere. They form in the freezing temperatures of the lower stratosphere, and are usually a mixture of nitric acid and ice crystals, sourced from parcels of moist air that are forced up through the tropopause – the atmospheric boundary separating the troposphere from the stratosphere – by the same orographic oscillations that are responsible for producing high lenticular wave clouds (see C_M4).

The likeliest time to see them is during a winter sunrise or sunset, when most of the sky is dark, leaving them lit by the sun from beneath the horizon. Their iridescent pastel colours can be magically beautiful.

There is a dark side to these iridescent wave clouds: their chemical composition assists the production of chlorine atoms, which in turn contributes to the depletion of the ozone layer.

The iridescent pastel colours of a nacreous cloud, photographed in the night sky.

BANNER CLOUD

Banner clouds, such as this impressive example, are caused by the physical presence of the mountain itself, which acts as an obstacle to the moisture-laden westerly wind, forcing it up towards the cloud-forming layer, just above the mountain's peak. Once this orographic cloud has been formed (orographic clouds are those created or influenced by mountains), it is pulled down the lee side by the reduced pressure on that side of the mountain, a motion that leads to the cloud's distinctive streaming pennant formation. The Matterhorn's banner clouds are the best-known examples of this phenomenon, although other large peaks, such as the rock of Gibraltar, create their own outstanding examples.

Banner cloud streaming westwards from the east face of the Matterhorn, Switzerland.

CATARACTAGENITUS

Clouds can develop in the vicinity of large waterfalls, from falling water broken up into ambient moisture. The plummeting cascade of water drags air down with it, causing neighbouring regions of moist air to lift up and replace it. This movement creates favourable cloud-forming conditions above the falls. Cataractagenitus clouds (the term derives from the Latin *cataracta*: 'waterfall'), which tend to be small in size, are given the name of the appropriate genus, followed by species or variety, plus the special cloud name, for example: cumulus humilis cataractagenitus or stratus fractus cataractagenitus.

A display of stratus fractus cataractagenitus, generated by the cascading waters of Niagara Falls.

SILVAGENITUS

Clouds often develop locally over forested areas as a result of increased humidity due to evaporation and evapotranspiration from the tree canopy. The process of forest cloud formation can often be seen on sunny mornings after rain, with spectral fingers of cloud appearing to rise and gather above the trees. Where these special clouds are observed, they are given the name of the appropriate genus and any appropriate species or variety, along with the special cloud name silvagenitus (from the Latin *silva*: 'forest'), as in the accompanying photograph of stratus silvagenitus.

Wisps of stratus silvagenitus appear over an area of forest following a bout of summer rain.

Man-made clouds

CONDENSATION TRAILS ('CONTRAILS')

Contrails are formed by the sudden condensation of the water vapour ejected from the exhaust of a jet aeroplane. At altitudes of 11km (35,000ft) or more, the outside temperature is well below freezing, so most contrails are formed of slowly sinking ice crystals. The opposite of a contrail is a distrail (short for 'dissipation trail'), and occurs when an aircraft flies through a natural cirrus or altostratus cloud, its vapour trail serving to overload the cloud with extra-heavy supercooled water droplets (or ice crystals), which then fall out, leaving a linear gap in its wake. If a contrail persists for ten minutes or more, it is classified as a special cloud, cirrus homogenitus (from the Latin for 'man-made'). In particularly moist upper air conditions, cirrus homogenitus clouds can persist for long periods of time and, under the influence of upper winds, spread for many kilometres across the sky. When this occurs, the resulting clouds are known as cirrus homomutatus.

Ice falling from an aircraft contrail (cirrus homogenitus).

Persistent aircraft contrails that spread across the sky are known as cirrus homomutatus. This image shows a mix of new and older contrails covering the sky near London City Airport.

FLAMMAGENITUS

Clouds can develop from convection initiated by heat from forest fires or volcanic eruptions. Such naturally occurring clouds are denoted by the special cloud term flammagenitus (from the Latin for 'fire-made'), such as cumulonimbus calvus flammagenitus, the large clouds that appear above erupting volcanoes.

Flammagenitus clouds were known until recently as 'pyrocumulus', a term that did not distinguish between clouds produced by natural or human activity, such as stubble-burning or other forms of deliberate combustion, as in the case of this low-lying, smoke-tinged cumulus humilis flammagenitus cloud hovering over a field of burning stubble.

The air needs to be fairly still for viable flammagenitus clouds to form, otherwise the thermal currents cooked up by the fire will be dispersed before the rising water vapour can reach the condensation point necessary for the formation of clouds. Once the fire is out, the cloud will soon decay and disappear, as would any other small cumuliform cloud when the convective currents that created it subside.

Cooked up from the thermal currents above a burning field of stubble, a pyrocumulus (cumulus humilis flammagenitus) cloud hovers in the evening sky.

CUMULUS HOMOGENITUS ('FUMULUS')

A variety of man-made cumulus cloud, 'fumulus' clouds typically form above industrial cooling towers. Much of the moisture that rises and condenses to form such clouds is emitted from the towers themselves over sustained periods, and this can combine with moisture already present in the atmosphere to produce significant and long-lasting cloud formations. These clouds are denoted by genus and species, followed by the special cloud name homogenitus (from the Latin for 'man-made'), in this case, cumulus mediocris homogenitus.

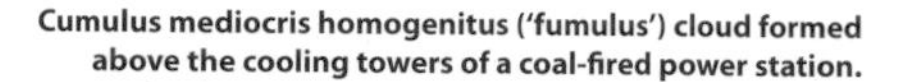

Cumulus mediocris homogenitus ('fumulus') cloud formed above the cooling towers of a coal-fired power station.

Optical phenomena and effects

HALO

A halo is an optical phenomenon that appears around the sun or moon when differently angled ice crystals in high cirrostratus or cirrus clouds reflect and refract the incoming light, splitting it into faint rainbow colours. Haloes are usually 22° in radius, and thus much larger than a typical corona (see page 129). Haloes often appear in conjunction with other optical effects, such as sun dogs and parhelic circles (see next page), and are often indicative of bad weather to come, 'haloes around the moon or sun means that rain will surely come', as an old weather saying succinctly puts it.

Haloes can also occur around other powerful light sources, such as streetlamps, as well as (in unusually cold conditions) above collections of minute ice crystals at ground level, known as 'diamond dust'.

A 22° halo around a streetlight, formed by the refraction of light through atmospheric ice crystals.

MOCK SUN / SUN DOG / PARHELION

Appearing as bright spots on either side of the sun, mock suns (also known as sun dogs, or parhelia) are another form of halo phenomenon, caused by the refraction of sunlight by ice crystals present in high cirriform clouds, especially cirrostratus. They tend to occur when the sun is low in the sky, and will appear to move closer to the sun the lower it sinks. Parhelia often exhibit a range of rainbow colours, with the red end of the spectrum appearing towards the inner edges, nearest the sun, and the bluish end appearing towards the outer ones.

It is the particular shape and orientation of the ice crystals within the cloud that determine whether sunlight will be refracted as a halo or a parhelion. Flat, horizontally aligned crystals will tend to produce parhelia, whereas more randomly shaped or poorly aligned crystals will tend to produce the 22° halo (see previous entry).

A sun dog appears alongside evening clouds, an effect caused by the refraction of sunlight by ice crystals high in the sky.

SUN PILLAR

A vertical streak or blade of light that appears above or below a setting or rising sun, sun pillars are produced by light reflected from the horizontal surfaces of slowly falling, hexagonally shaped ice crystals that are present in cirriform clouds, especially cirrostratus clouds (see C_H7, page 90). Unlike refraction phenomena such as haloes and sun dogs, pillars are simply the collective glittering of millions of tiny icy crystals, and will thus reflect the colours of the surrounding sunset, rather than divide into the spectrum colours that are associated with the process of refraction.

On rare occasions, a sun pillar will coincide with a parhelic arc, forming a bright cross in the sky, centred on the sun.

A sun pillar appears at sunset on a cold February evening in Latvia.

CIRCUMZENITHAL ARC

Caused by the refraction of sunlight through the same horizontally aligned cirriform ice crystals that are responsible for producing sun dogs (see page 126), a circumzenithal arc (CZA) is a band of bright prismatic colours that resembles an inverted rainbow, positioned immediately above the viewer's head.

Typically a quarter-circle in shape, its colours, which are often brighter than those of the rainbow, run from blue near the zenith, down to red near the horizon. Due to the precise angle in which the refracted light exits the sides of the horizontal ice crystals, CZAs cannot occur if the sun is more than 32.2° above the horizon, and the brightest arcs of all occur when the sun is exactly 22° above the horizon.

An ice rainbow, known as a circumzenithal arc, makes an appearance in the early evening sky.

CORONA

A corona is a coloured ring or sequence of rings surrounding the sun or moon, as seen through thin layers of low or medium cloud. As can be seen in this impressive example of a corona in altocumulus, the colours of the corona tend to be bluish or whitish nearest the centre, shading towards the red end of the spectrum further out. As coronas are caused principally by the diffraction (bending and scattering) of light by uniformly sized water droplets, the phenomenon is closely related to that of irisation.

A rarely seen form of the corona is the Bishop's ring effect, caused by dust and sulphate particles lingering in the atmosphere following major volcanic eruptions. The effect (named after Sereno Bishop, who first described the phenomenon in the wake of the eruption of Krakatau in August 1883), is of a wide corona around the sun, the inner rim of which is whitish or bluish in colour, while the outer part tends to reddish, and even purplish.

A bright corona appearing through a layer of altocumulus clouds.

IRISATION

Irisation is a form of light interference caused by uniformly sized water droplets serving to diffract sunlight, bending the light round the drops rather than allowing it to pass straight through. The light typically recombines to form irregular patches of luminous pastel or mother-of-pearl shades, predominantly green and pink, at the edges of thin cirrocumulus, altocumulus and stratocumulus clouds. The word is derived from Iris, the Greek goddess of the rainbow.

Irisation in altocumulus clouds.

GLORY

A glory is an optical phenomenon produced by light scattered back towards its source by a cloud of uniformly sized water droplets. Appearing as a sequence of coloured rings, glories are sometimes seen by mountaineers in association with the so-called Brocken spectre, the magnified shadow of a person that is cast by low sunlight on to the upper surfaces of clouds that are situated below where the viewer is standing. The Brocken spectre is named after the cloud-shrouded peak in the Harz mountains of northern Germany, where the effect was first recorded.

Glories are now most often observed from aeroplanes, as in this example, taken from the window of a plane.

RAINBOW

Rainbows are caused by water droplets dispersing sunlight into banded arcs made up of the seven colours of the visible spectrum: red, orange, yellow, green, blue, indigo and violet. White light is refracted as it enters the droplet, then reflected off the back surface, and re-refracted on the way out, being thereby dispersed in a wide range of angles, the most visually intense of which will always be between 40° and 42°, due to the spherical nature of the droplets. Rainbows can only be seen when the sun is behind the observer, and the airborne water droplets – whether from falling rain, spray from a waterfall, or even from a garden sprinkler – are directly in front. Being entirely optical phenomena (rainbows do not exist in any tangible sense), the eye of every

Primary rainbow.

Primary and secondary rainbow.

observer will constitute its own individual rainbow, with a unique position in the sky that is always exactly opposite to the sun with respect to where each individual observer happens to be standing.

The most commonly seen bow is the primary bow, which is always 42° in radius, with red on the outer edge and violet on the inner edge. Sometimes a larger secondary bow (52° radius) accompanies the first, as can just be seen in the top of the photograph above, in which the colours are reversed, red on the inside, violet on the outside. Pale arcs, known as supernumerary bows, also sometimes appear inside the primary bow.

CREPUSCULAR RAYS

Beams of sunlight that are scattered and made visible by minute particles within the lower atmosphere, especially dust, gases and water droplets. There are three recognised varieties of ray, their differences due to variations in atmospheric transparency – 1: rays which emerge from gaps in low cloud, caused by the scattering of light by water droplets; 2: beams of sunlight which rise from behind a cloud (usually a cumuliform cloud); and 3: pinkish rays which radiate from below the horizon, their intense coloration caused by the presence of haze. The downward streaming 'Jacob's Ladder' variety, as seen here, takes the form of columns of sunlit air, separated by shadows cast by the intervening cloud, which appear to converge on the sun.

Crepuscular rays, 'Jacob's Ladder' variety.

AURORA

Appearing as tall ribbons of multicoloured light, or as more diffuse glowing patches in the sky, *c.*100 to 250km (60 to 150 miles) above polar latitudes, aurorae are produced by fast-moving particles from the streaming solar wind as they collide with atmospheric gases in the Earth's upper atmosphere, causing gas molecules to fluoresce in a range of different colours. Aurorae can be seen at both poles, and are known as *aurora borealis* (or 'The Northern Lights') in the north, and as *aurora australis* in the south. They are named after Aurora, the Roman goddess of the dawn.

The Northern Lights: formed as the solar wind collides with Earth's gaseous atmosphere.

LIGHTNING

A visible electrical discharge produced by the actions of cumulonimbus thunderclouds, lightning is formed when pockets of positive (+) and negative (-) charges within an individual cloud become separated by the action of violent updraughts of air. Clusters of electrical charge will accumulate in different parts of the cloud, with positive charges often congregating at the top, and negative charges at the base of the cloud. When the potential difference between these charged areas becomes too great to sustain, electrical energy is discharged in the form of lightning.

Lightning strikes the ground when the negative charges in the base of the cloud induce opposite charges in the ground below, sending a powerful spark flying between the differently charged regions in the form of cloud-to-ground lightning (often described as forked lightning). Sparks that fly between differently charged regions within an individual cloud, and thus do not reach the ground, are known as in-cloud (or sheet) lightning, while sparks that fly across the air from one charged cloud to another are known as cloud-to-cloud lightning.

Cloud-to-ground lightning.

Lightning often seeks the ground in 'steps', as can be seen in this photograph of forked lightning striking the Earth.

Glossary

accessory cloud: a supplementary cloud that occurs in conjunction with one of the principal types; they are pannus, pileus, velum, and flumen. There are also eleven supplementary features which may also appear: incus, mamma, virga, cavum, praecipitatio, arcus, murus, cauda, asperitas, fluctus, and tuba.
altocumulus (Ac): medium-level clouds, occurring as individual rounded masses, often with clear sky visible between them.
altostratus (As): medium-level, dull white or blueish layers of cloud, which tend not to produce rain.
arcus (arc): from the Latin for 'arch' or 'bow', an arcus is an accessory cloud in the form of an arch.
asperitas (asp): from the Latin for 'roughness', asperitas is a wave-like supplementary feature that appears on the underside of stratocumulus and altocumulus clouds.
atmosphere: a c.500-km (c. 300-mile) band of gases encircling the Earth, made up of five layers, each with its own distinct temperature profile: the troposphere, the stratosphere, the mesosphere, the thermosphere, and the exosphere.
calvus (cal): smooth or bald-topped clouds, in contrast to the hairy or striated appearance of capillatus clouds (see C_L3).
capillatus (cap): a striated anvil of ice crystals that appears only above cumulonimbus capillatus clouds.
castellanus (cas): from the Latin meaning 'castle-like', these clouds exhibit cauliflower-shaped turrets at their summits.
cataractagenitus (cagen): a special cloud that forms in the spray above large waterfalls.
cauda (cau): a horizontal, tail-shaped structure that extends from the main precipitation region of a cumulonimbus storm cloud.
cavum (cav): also known as a 'fallstreak hole', this supplementary feature appears when a patch of supercooled cloud freezes and falls away.
cirrocumulus (Cc): thin, white patches or layers of high cloud. A large-scale display of cirrocumulus is known as a 'mackerel sky'.

cirrostratus (Cs): transparent, whitish veils of smooth or fibrous cloud, made up of layers of ice crystals high in the sky.
cirrus (Ci): high, wispy clouds, often with a fibrous or silky sheen. They are composed of millions of slowly falling ice crystals.
cloud: a collection of minute particles of liquid water or ice, suspended in the air and usually not touching the ground. It may also contain certain non-aqueous liquids, as well as small solid particulates such as salt-grains, pollens, smoke or dust.
condensation nuclei: minute airborne particles of dust or other solid material on which water vapour condenses into droplets; their presence is necessary for the formation of clouds.
congestus (con): tall, thin clouds with cauliflower-shaped tops.
cumulogenitus (cugen): clouds formed from the spread or decay of cumulus clouds.
cumulonimbus (Cb): convective Cumulonimbus clouds (C_L3 and C_L9) can grow to immense heights, producing lightning, hail, and stormy conditions at ground level.
cumulus (Cu): detached, dense clouds, the upper parts of which appear brilliant white in sunshine.
duplicatus (du): from the Latin for 'doubled' or 'repeated', clouds of this variety persist in more than one layer.
fibratus (fib): straight clouds, with no hooks (as distinct from clouds of the species uncinus).
flammagenitus (flgen): clouds that form over a direct source of heat, such as a forest fire or erupting volcano (also known as 'pyrocumulus').
floccus (flo): from the Latin for 'tuft of wool', clouds of this species are small and tufty, often with ragged lower parts.
fluctus (flu): distinctive wave-like crests that form above a variety of high and mid-level clouds. Also known as 'Kelvin–Helmholtz waves'.
flumen (flm): from the Latin for 'flowing', flumen are bands of low accessory clouds associated with severe convective cumulonimbus storm clouds.
fractus (fra): from the Latin for 'broken', clouds of the species fractus are ragged, sometimes patchy in appearance.

homogenitus (hogen): from the Latin for 'man-made', the term applies to any clouds that arise from human activity, such as aircraft contrails (cirrus homogenitus) or the clouds that appear above industrial cooling towers (cumulus homogenitus).
homomutatus (homut): from the Latin *homo* ('man') and *mutatus* ('changed'), the term refers specifically to persistent aircraft contrails that have spread to cover much of the sky.
humilis (hum): small, flattened clouds that are generally wider than they are tall.
incus (inc): from the Latin for 'anvil', an incus is an icy canopy that forms at the summit of a cumulonimbus capillatus cloud.
intortus (in): from the Latin for 'twisted', clouds of this variety are irregularly curved or twisted in appearance.
irisation: an optical effect where rainbow colours appear at the edges of cloud as they pass across the sun or moon.
lacunosus (la): from the Latin for 'with gaps or holes', clouds of this variety are reticulated like a net.
lenticularis (len): almond or lens-shaped wave clouds, often formed by the movement of moisture-laden air over a hill or mountain slope.
mamma (mam): distinctive pouches hanging down from the under surfaces of cumulonimbus clouds.
mediocris (med): clouds of this species are generally of equal width and depth, often with small bulges at the top.
murus (mur): a large, localised lowering of cloud that develops beneath the surrounding base of a cumulonimbus storm cloud. Also known as a 'wall cloud'.
nacreous clouds: icy, pastel-coloured clouds that form in the lower stratosphere.
nebulosus (neb): clouds of this species are thin, misty or veiled in appearance.
nimbostratus (Ns): dense, grey blankets of cloud from which drizzle or persistent rain often falls.
noctilucent clouds (NLC): thin, icy clouds occurring high in the mesosphere.
opacus (op): dense clouds that completely mask light from the sun or moon.

orographic: created or shaped by the presence of high mountains.
pannus (pan): accessory clouds in the form of ragged shreds that usually appear below rain clouds.
perlucidus (pe): clouds of this variety allow some sun or moonlight to be seen.
pileus (pil): a flattened, cap-shaped accessory cloud, occurring mostly with cumulus and cumulonimbus clouds.
praecipitatio (pra) : precipitation of any kind (i.e. rain, snow or hail) that reaches the ground (as distinct from virga).
radiatus (ra): parallel bands or rays of cloud which appear to converge.
silvagenitus (sigen): special clouds created by the evaporation of moisture above large forested areas.
spissatus (spi): dense clouds, often grey in colour.
stratiformis (str): wide sheets of cloud that extend horizontally.
stratocumulus (Sc): rounded masses or rolls of cloud which often appear in parallel bands.
stratus (St): low, sometimes indistinct layers, which seldom produce rain.
translucidus (tr): clouds of this variety show sun- or moonlight clearly through.
tuba (tub): funnel clouds which extend downwards from the base of a cumulonimbus cloud, but rarely reach the ground.
uncinus (unc): comma or hook-shaped.
undulatus (un): clouds of this variety feature parallel waves or undulations.
velum (vel): an accessory cloud of great horizontal extent, in the form of a veil enclosing the upper part of one or more cumuliform clouds.
vertebratus (ve): clouds of this variety (typically cirrus) appear like ribs or fishbones in the sky.
virga (vir): streaks of precipitation (usually rain or snow) which fail to reach the ground.
volutus (vol): a detached, tube-shaped cloud mass, that often appears to roll slowly about a horizontal axis. Appears mostly with stratocumulus and occasionally altocumulus.

Index

A DAVID AND CHARLES BOOK

David and Charles is an imprint of David and Charles, Ltd, Suite A, Tourism House, Pynes Hill, Exeter, EX2 5WS

First published in the UK and USA in 2008
This updated edition first published in 2023

A catalogue record for this book is available from the British Library.

ISBN-13: 9781446310113 hardback
ISBN-13: 9781446313152 EPUB
ISBN-13: 9781446313169 PDF

Printed in China through Asia Pacific Offset for: David and Charles, Ltd, Suite A, Tourism House, Pynes Hill, Exeter, EX2 5WS

10 9 8 7 6 5 4 3 2 1

Publishing Director: Ame Verso
Senior Commissioning Editors:
Freya Dangerfield and Lizzie Kaye
Managing Editor: Jeni Chown
Editors: Emily Pitcher, Susan Pitcher, James Brooks and Sarah Tempest
Design Manager: Anna Wade
Designer: Jodie Lystor
Pre-press Designer: Ali Stark
Production Manager: Beverley Richardson

Layout of the digital edition of this book may vary depending on reader hardware and display settings.

PICTURE CREDITS

Cover © Unsplash/sam; 2 © Unsplash/billy_huy; 4 © Unsplash/sendi_r_gibran; 7 © S Mallon; 9 © V&A Images; 14, 16, 17, 33 (bottom), 35, 42, 66, 74, 110, 111, 118, 119, 127, 129, 130 © Wikimedia Commons; 18 © J Bell; 14, 20 © S Jebson; 14, 21 © N Goodban; 15, 23, 73 © M Clark; 25, 30, 47 49, 59, 105, 106, 107, 113, 132, 133 © R Coulam; 14, 15, 29, 51, 54, 60, 62, 63, 71, 79, 83, 87, 88, 89, 91 © CS Broomfield; 14, 15, 27, 30, 31, 40, 67, 72, 85, 92, 94 (top) © RK Pilsbury; 14, 32, 33 (top), 37, 45, 61, 98, 102, 120 © JFP Galvin; 36 © WG Pendleton; 14, 38 © A Bushell; 14, 15, 41, 44, 56, 69, 70, 97, 104, 112, 134, 137 (both) © Crown Copyright 2021, Met Office; 14, 46 © JW Warwicker; 14, 50 © R Ham; 14, 15, 52, 64 © KE Woodley; 14, 15, 55, 58 © C Irving; 57, 84, 95 (bottom), 99 © J Corey; 65 © SG Cornford; 75 © JM Pottie; 15, 77, 131 © A Best; 15, 80 © M Kidds; 15, 82, 125 © J Walton; 90 © RW Mason; 93 © WS Pike; 15, 96 © RD Whyman; 100 © A Simpson; 101 © Steve Willington; 103 © PJB Nye; 109 © NOAA Photo Library, NOAA Central Library; OAR/ERL/National Severe Storms Laboratory (NSSL); 115 © M. Yrjölä; 116 © PJB Nye; 117 © MJO Dutton; 121 © R Hamblyn; 123 © GJ Jenkins; 124 © CG Holmes; 126 © R Stagg; 128 © R Selby; 135 © DAR Simmons.